BURNING SNOWFLAKE

by the same author

HAWESWATER
THE ELECTRIC MICHELANGELO
THE CARHULLAN ARMY
HOW TO PAINT A DEAD MAN
THE BEAUTIFUL INDIFFERENCE
THE WOLF BORDER
MADAME ZERO
SUDDEN TRAVELLER
BURNTCOAT
HELM

as co-editor
SEX AND DEATH

SARAH HALL

Burning Snowflake

NATURE · WRITING · REBELLION

FABER LECTURE

DELIVERED IN MANCHESTER AND LONDON

2026

faber

First published in 2026
by Faber & Faber Limited
The Bindery, 51 Hatton Garden
London EC1N 8HN

Typeset by Faber
Printed and bound by CPI Group (UK) Ltd, Croydon CR0 4YY

A CIP record for this book is available from the British Library

ISBN 978–0–571–40462–9

MIX
Paper | Supporting
responsible forestry
FSC® C013604
www.fsc.org

Printed and bound in the UK on FSC® certified paper in line with our continuing commitment to ethical business practices, sustainability and the environment.
For further information see faber.co.uk/environmental-policy

Our authorised representative in the EU for product safety is
Easy Access System Europe, Mustamäe tee 50, 10621 Tallinn, Estonia
gpsr.requests@easproject.com

2 4 6 8 10 9 7 5 3 1

BURNING SNOWFLAKE

Dear Reader,
I am trying to pry open your casket
with this burning snowflake.

— James Tate

I was born and raised beside a river, in an old whitewashed cottage surrounded by fells.

If this sounds like the beginning of a nostalgic Lakeland memoir, or the first line of an upland folk ballad, don't worry, it's neither. But I do want to talk about my journey as a writer, and I want to talk about nature, how nature is written, and how it has become our most crucial story – upon which the future depends.

Nature stories are surely the oldest and most foundational form of storytelling. Worldwide, origin myths begin with the making of continents, mountains, elements and species. Eschatology imagines the Earth and its inhabitants destructed: natural paradise is staged by God, then lost to humans. Science, too, models the evolution of the cosmos – an expansive story, starting with a very hot moment and ending with tearing, recompression or a chilled subatomic soup. Narratives are also key to our current environmental predicament – the comprehension, apprehension and politicisation of climate crisis and Earth systems breakdown. Data, speculation, denial, adaptation, survival, extinction – there are many narrative perspectives and presentations of the situation, some convincing, others less so. Nature writing today faces huge challenges: it must deal with this complexity, this existential reckoning.

But I should start with my own beginning. I'm often asked about my upbringing and the relationship of landscape to my work. So often am I asked that my answers have become overtrodden and worn-sounding, to me at least. This is the Cumbria conundrum, an inescapable tie between literary profile and content, an association both pride-inducing and annoyingly restrictive. The Lake District is, of course, famed for its scenery and symbols; it has been branded by the art and literature of previous eras – gents whose surnames begin with W. Picture painterly hills and pencilled fells, lonely rain clouds, and smiley, goggle-eyed Herdwick sheep printed on overpriced mugs. Were he around now, no one would ask Wordsworth whether he was inspired by New York City or the sartorial fashions of the day (the switch from breeches to pantaloons, if you're wondering). No one would ask Wainwright whether his mountain sketches were based on an appreciation of obscure French cinema.

My usual reply to the question is that the natal influence of landscape was so fundamental that it has formed the DNA of my writing. It goes deeper than inspiration or muse, sensibility or legacy; it's more programmatic and organismic. Place built my imagination. This declaration sounds rather profound, if vague and biologically unverifiable, but it does convey an honest belief, some summary knowledge of the self. I should interrogate the relationship properly, though – the language of nature and the nature of language – if the alchemical connectedness forms my artistic base. And I should ask what this operating system means in these turbulent, ecocentric times.

So. I was born and raised beside a river, in an old whitewashed cottage surrounded by fells.

The cottage sat above the lovely watercourse of the Lowther Valley, in the district of Eden – a name not insignificant to the young impressionable brain. The area was quite beautiful, but it was not the primordial, delightful paradise of biblical base camp. The valley was in the historic county of Westmorland, which was amalgamated with Cumberland in 1974, the year of my birth. It was remote, with working villages and hamlets, ancient upland farms, and the Haweswater Dam at its southern head, the sluice gates of which could let substantial volume into the river, leading to daydreams about dramatic breaches, tidal waves, disaster-movie catastrophes. Our garden gate led out onto scruffy, thistle-strewn banks and open moorland, what seemed in childhood to be wilderness. My bedroom window faced the High Street mountain range, which divided the valley from Ullswater and the Lake District proper. On our side was Realism: utility, school, agriculture, an eerie, mountain-ringed reservoir that served Manchester; the odd lost traveller or film director looking for a terrible holiday venue for *Withnail and I*. On the other side was Romanticism: crowds of cagouled visitors, a national park with baggable peaks, conceptual territory as well as relief-mapped. They had Willie and Dorothy, lake steamers and gingerbread; we had the last golden eagle in England, carlin pea suppers, and carriage-driving trials at which the Duke of Edinburgh was trounced annually by our local horse-breeding family. (All hail the People's Republic of Cumbria.)

I mentioned that Westmorland vanished in the year of my birth. This fact doesn't matter, other than as an example of how shifting the region has been, how alterable its lines, its names and identities. The north of England was once known as the Debatable Lands. It still is, some might argue, and I'd probably agree. The borderland was claimed and disputed by nations, clans and families, bandits and reivers; it was feudal and violent, lawless for centuries – the original government-managed decline. It was a staging ground for conflict and rebellion, for cattle rustling, football games played with the heads of Englishmen, Scottish incursions repelled by the arrows of the Kendal bowmen. Not exactly a pastoral idyll. Cumbria's drystone walls are Viking. Soldiers from as far afield as Syria were stationed along Hadrian's Wall, a structure marking the halt-point of an empire predicated on expansion. The territory has been home to visionary redefinitions – transformed from pointless, untamed wasteland into a sublime and beautiful wonderland, 'Wordsworthshire' – thanks to those chaps in heroic frock coats and stout walking boots. Then converted back into barren wasteland by overgrazing – a 'sheep-wrecked' wildlife desert. Birthplace of modern conservation; crucible for the fervent arguments around rewilding. Over the centuries, my home turf has been feared, despised and valorised, settled, abandoned and repurposed. It is, consequently, rich with oppositional history, and primed for possibilities.

When I was growing up, I wasn't aware of what this dissonance meant, if anything. I was too busy being a wildling, climbing trees and hills, making dens, absolutely river-obsessed. The river was the best thing a childhood could have. I loved being in it, no matter the temperature or the state of the water – sometimes clear as glass and burbling away, sometimes dangerously in spate and ferrying dead livestock, sometimes thickly coated with snot-like toxic run-off. I loved building rafts from planks and ropes and tractor tyres with my older brother, imagining we could boat down our tributary to the Eamont, then to the mighty River Eden, even as far as the Solway Firth.

But much of the time I was alone with the water. I loved sitting on boulders in the current, scanning for trout and eels, an occasional salmon trying to migrate up the straightened streamway to its vanished redds. Catching minnows with a net for my jam jar, gently lifting rocks to find crayfish huddled underneath. For hours, I would study the river's gorgeous, sensual worlds, its endless formless story. It constantly, mesmerically described itself to me. It was the voice of my thoughts, lyric, a limerence.

And I knew the river was somehow making me up too; that I was being described by it. Interiority, exteriority. Body and consciousness. The more solitary I was, the more I merged with the world. I was not apart from the river, I was permeable, and there was no need for the writer's act of transference – atom to simulacrum via words. Every day I walked along the river's banks, upstream past sand martin nests and otter holts, the hares' ears flicking dark Vs at me above the tawny moorland grass. Or I walked down through mossy woods towards the Lowther Estate, where there was a wildlife park containing deer, boar, peacocks, snakes, exotic owls. And wolves. Sometimes I dreamt the wolves were not in their fenced enclosure but running free through corridors of trees and across grassland, past our cottage and away over the horizon.

My parents had relocated from the London commuter belt in the early seventies, when my brother was six weeks old. They were done with the city's grind, the crowds and crushes; they wanted fresh air, greenery. The valley was affordable then, not prime location. Within a few years my dad got the job he would hold for three decades, in a paper mill; he had to travel over Shap Fell every day to Kendal, via the highest road in England. My mum couldn't drive then – to get the shopping we would walk over the moors and fields, fording the river in the shallows or using the 'swing bridge', a construction of rusty metal cables and old wooden slats, which bounced and shook and swayed under the weight of anyone crossing. We'd navigate bulls and rams en route, floodwater, mud, carcasses. Then we'd pick up the service bus to Penrith, go to the old covered market and the butcher, and to the library. When Mum and Dad first moved into the cottage, there was no heating, just two open fires; there were jackdaws in the roof and shrews in the walls, and when the nearby lambs bleated they thought it was my baby brother crying.

All around the house were flocks and herds, semi-feral ponies, industries and traditions not dissimilar to those of a hundred years before, two hundred. The weather was big and instructive: blizzards, unending rain, windstorms and intermittently scorching summers – one so desiccating that the Haweswater Reservoir shrank and exposed the skeleton of the drowned village of Mardale, its deconstructed houses, hotel and church, its perfectly intact humpback bridge. Everyone adapted around the seasons and the calendar, which meant hard labour, arrangements with the land, with the climate and with neighbours. The valley was not ethnically or racially diverse: there was a Polish refugee, our district nurse was from the West Indies and my godfather was Jamaican, by way of Canada. The population was small; out and about, I was more likely to encounter an animal or a bird than I was another human being, and that in its way was companionable, an influential demographic. Locals were succinct, wry, earnest and very resilient – the women especially; a woman with a gun or a sheep across her shoulders, manoeuvring big machinery or upbraiding her fella for his clarty boots, was not unusual. Nor was hearing Westmerian dialect, at agricultural events, in the pub and the post office, and at school. There were rustic celebrations, sports days and vintage steam gatherings, harvest festivals, church services and young farmers' dances – ways to lighten the dark, to leaven heavy lives and lift spirits. I remember Bonfire Nights in the freezing November sleet, sparks colliding with ice, snow impossibly on fire. I remember hot soup served from metal horse trailers, carolling at the highest fell farms in deep drifts, torching the gorse at Halloween and jumping over burning bushes. I remember hollying, home-brewing and outdoor wrestling – there were prizes for the best floral-embroidered wrestling suit as well as for champion hipes and hanks. It wasn't privileged or paradise, it was just rural. Perhaps, when you're undergoing exposure, it doesn't always register as a state of being. That comes later.

This isn't a memoir, but my point is: it was my habitat, my home, my form of existence. My parents were offcomers and didn't quite belong. But I was born to the north-west quarter, and I did. The river raised my consciousness, taught me implicity, told me who I was and what we were. How has landscape made me the writer I am? I believe no impedimentary boundaries exist between this world and the mind's world and the world we create. Perhaps that sounds mystical, transcendental; it isn't meant to. I simply believe the realms are porous, reconstructable, and each is a version of truth. I am a nature writer, because I am, inseparably, nature.

When we think of nature writing, we may not immediately associate it with radical players and cutting-edge discourse, or picture an author in warrior pose. We might recall the cultivated, observational prose of school English lessons:

poetic arcana about foxes' tails and frozen ponds and fifty shades of moss. We might think of naturalists' rhapsodies about beach pebbles and adamantine odes to the wind, Latinate field guides and botanical encyclopaedias. An institutional, predominantly white male corpus. There have been very provocative nature writers in the UK, of course, those for whom the countryside was a political arena, those writing about access, inequity and poverty. John Clare's nature was not metaphorical or parochially kiltered, but a destructive site where enclosure was ruinous and the poor were disenfranchised by it, stripped of their rights to land. R. S. Thomas's Welsh nationalism was enacted through the harsh symbolism and geology of his verse. More recently, the poems and essays of Kathleen Jamie question the traditional dominant narratives about place and the position of the anthropocentric writer in the wild; her conversations with nature prefer biocentrism, a narrator no more or less important than birds, animals, fauna, the stone of a Highland cherry.

Nature writing is a broad, encompassing term now, and has boomed in Britain, as demonstrated by the colourfully arrayed piles on the tabletops of bookshops. Its inclusivity means it is hard to categorise, with wide-ranging themes and styles, including personal reflections, cultural histories, topographies, folklore, travelogues and ecopolitics. But its core tends towards non-fiction. Accounts about curative involvement and unlikely relationships with wildlife, such as Helen Macdonald's *H Is for Hawk* and Chloe Dalton's *Raising Hare*, are runaway successes. As are the gorgeously illustrated foraging manuals and 'wild swim' guides. As are the adventure yomps into the Arctic or the Amazon, or around the coast of Britain. At the sharper end of the spectrum are the environmental treatises, such as Robert Macfarlane's *Is a River Alive?*, which calls for legal protection of nature. George Monbiot's work goes further, examining how consumer capitalism, neoliberalism and land ownership are the underlying tectonic plates affecting the Earth's welfare, while Anita Sethi highlights the historical racial zoning of the countryside and asserts the right for all people to belong there, for the land to belong to all people. Nature writing is flourishing, to the point of mass marketability, a lifestyle sell. Such books might be both coffee-table grade in aesthetic and the activist's accomplice.

So where do I fit on the shelf? Notwithstanding this offering, I write fiction. Fiction is, essentially, flim-flam. An elaborate, interactive fabrication of setting, character, plot and theme. Virtual reality rather than terra firma, artificial drama rather than reportage. Where does fiction fit in the lexicon of nature writing, and in the paramount conversation about the health of our natural world? What can fiction writers contribute to the most urgent subject of today with their conceits and concoctions, their diverting jackanories?

To answer, I need to return home again. To the influence of landscape on my writing, and that formative immersion. To the *where* of the imagination. Novelists often talk about character being the driver for a new work, a way to connect with the reader empathically. But place is, and always has been, the primary element of composition for me. When I begin a story, I first think of its location. I think of the necessary literary building blocks of wherever *where* is, be that Morecambe Bay, Mozambique, Finland or Brooklyn. The populous. The psychogeography. The angle I must take in order to construct a place, unfamiliarity or intimacy, knowledge and bluff, a crux between. This is true for historical fiction and speculative futurism too, all stations in time. I must capture the visual details, the sounds and smells, the verdure or urbanity, the architecture, residents, signage, vibe-age, the elements. I must think hard about ways to describe *where*, the language employed in its illusion and evocation. Which words best communicate the essential qualities, the physics and soul of a place; which words create a beguiling rendition and reader interface. I owe this squarely to my upbringing – the compulsion to craft living, vivified, experiential settings; settings commensurate to the visceral, aromatic redolence of Cumbria. That In-Worldness. Geographic veracity has to be transferable, because a city flower must be attended to as respectfully as the Pacific or Helvellyn.

Take my river. What I want is to decant it from my experience and incant it into the reader's. What I want is for the reader to believe utterly in the river on the page. To see it and hear it and feel it. To smell and even taste it. Otherwise, what's the point? If you don't believe you are there, if you aren't convincingly transported to the water's edge, or over, or into its flow, surely nothing else within the fictive domain can be 'true' either, no holistic suspension of disbelief possible.

Please allow me a moment of flim-flammery; I just can't help myself.

Here it is. A supple artery in the heart of the valley. Silver and animate. Sky-reflecting, reddened by the bloom of sunset. There's light glinting off its surface – cymbals of light, light tilting and sledging in the current, jinks of light breaking in the shallows and exploding in the rapids, light like scales along sinews, light expanding in rings where fish mouth the surface, shellacked where the river is impounded in pools and polished. Here is the soft, cold fetch of water moving past. Clear volume freighted seawards, languid in the straights, pouring faster over declivities. Rushing over cobbles and around reeded islets, glupping between rocks, combing beards of weed downstream. Torrenting brownly after rain and spilling across fields, intimidatingly silent as it surges over the banks. Carrying its corrupt, complicated fragrance, of minerals and organisms, silt and peat, skunked matter and silage. The river's scent is fluvial, and almost female, venal, sexual. Stones glint in the bed, amber, slate, sorrel, ochre, their colours vernacular, activated by moisture, made sacred by the river's own spirit.

Why try so hard forging phrases and sentences, using them to conjure and persuade, trying to pry open the reader's casket? Why employ such a strange, paradoxical practice, using the artifice of depiction to manufacture a believable premise? Because we can travel anywhere together via the right words – to the moon, the poles, the tropics, to old-growth forest and submarine canyons, to villages, towns and metropolises, to pre-human history, timelines of civilisations, to post-Anthropocene ruins or new adaptive compounds, Mars farms. All the iterations and projections of our environments. All the versions that were, that are, and possibly might be. We can get inside the king's bedroom, or a missile silo. Land in the Andaman Islands, enter the Apostolic Archives, ascend to La Rinconada. Through the portal of prose we are granted freedom to pass, to trespass and to access multiverses, their conditions and climates, their experiences and influences, their effects, perhaps, on our own location when we close the book and arrive back home. *Where* becomes *wherein*.

If you grow up in the Lake District and you're becoming a writer, at some point there will likely be a negotiation with Romanticism. If it's not artistic, it will be associative. Cumbria is a county full of artists and artisans, with a history of visions drawn from and inspired by nature. It's a place that's creatively catalytic, a literary goldmine. But there's an arrested development to the District's legacy, about which a writer today might feel very ambivalent. Wordsworth and Romanticism still prevail, and are intrinsically linked to the region's UNESCO World Heritage status. The poet's idealised 'perfect republic of shepherds' has helped shape the idea of cultural uplands, as well as actual physical ones, granting protected status to a farming practice that is, today, very different from the small-scale democratic yeomanry it sought to celebrate; that is, in fact, loss-making, heavily subsidised, and at current stock numbers environmentally ruinous. Writers such as James and Helen Rebanks have argued with nuance and pragmatism for the preservation of Cumbria's unique sheep trade and way of life, while conservationists try to balance that against biodiversity loss and erosion. With the UNESCO listing there's a conscious injunction to cherish a dated idea *of* the place and the ongoing influence of this idea *on* the place. It's quite the mantle.

I'm perhaps suggesting I have never been a signatory, that I've always been free of the Romantic legacy, but when I wrote my first novel I wasn't. *Haweswater* is historical fiction, set in the interwar years, which, given the remoteness and isolation of the area, harks back to an even older era. It's a story about the impact of industrialisation on a hill-farming community; the building of a dam, the displacement of people who have lived in a remote dale for generations. I grew up 'under the shadow' of the Haweswater Dam, and in my village were former residents of Mardale. The novel is based on a true story and its primary theme is never out of fashion. Seldom does big industry not displace or disadvantage a vulnerable, marginal section of society, even when it's pitched as being for the greater good. My version is elegiac and deeply bucolic, a hymn to a lost populace. It demonstrates a keenness to be accurate, topographically and historically, to honour the past, the map and the old farming ways. Within this book are fizzing, first-publication nerves, a desire to be authentic and not piss off lovers of the Lake District. I partially corseted myself in my debut, to fit with the expectations and soft politics of an iconic territory.

When *Haweswater* was published, I received Hardy and Lawrence comparisons. This perhaps reveals how few intervening rural novels had made it up the ladder and also how few women novelists have been lauded to canonical standards for them. I was an outsider, writing from and about the 'backwaters', far from the centrifuge of London publishing.

Haweswater was off-register rather than avant-garde, and though it was granted the generic and fairly depreciative label of 'regional writing', its highlighting of the common labourer's plight through the fiery, dissenting farmer's daughter, Janet, attracted readers around the world. It is the story of many tenanted agricultural communities.

Such creative inheritance is weighty, and ironic, given that one of the central tenets of the Romantic movement was the concept of mutability: change being the only constant. Territory does change, politically, demographically and, most importantly, ecologically – even territory where landscape is being preserved by nostalgic aspic, heritage metrics and large grants. Quite quickly, I realised that what had actually been handed down to me from my literary antecedents was a protean notion about flux and adaptation. Nature is always evolving, so we must evolve with it, and as it. And fiction derived from nature can represent and engender these changes. This isn't the biscuit-tin principle of Romanticism sold in the villages of the District today, but a more integrous principle of it. I'll return to these seeds of resistance and revolt later.

In some ways, I did not realise just how free I was to write *Haweswater*. Looking back to the period when it was drafted, three decades ago, I see myself with absolute liberty to choose environmental subject matter, and follow classic pastoral traditions. Those traditions might feature tropes of city versus country, and nature versus industry, but I felt no pressure to respond to contemporary climate discourse. Society was aware of issues such as ozone depletion, deforestation and pollution, but the systematic totality and urgency of our human impact on the planet was yet to become the elephant in the nature writer's room. I was simply grateful for the opportunity to bring my literary preoccupations to readers. I knew then, as I know now, it's a privileged platform to hold. That green and pleasant stage wouldn't remain so for long, though – how could it, given all we were learning about the threat to our habitat and atmosphere, the uncomfortably mounting evidence?

Within a few years of *Haweswater* being published, a wave of alarming non-fiction books arrived in the marketplace, scientifically researched and predictive works mapping the clear dangers of fossil-fuel consumption, biodiversity loss and temperature rises. *Six Degrees* by Mark Lynas tracked the escalatory and deadly results of global warming. *Half Gone* by Jeremy Leggett, a former geologist, exposed the impending disaster scenarios of the peak oil phenomenon. These apocalyptic primers were utterly terrifying. And the issues were not simply speculative or down-road; they were already live in the world. The sad and solipsistic truth about climate engagement is that until it happens to you, to your home, to your family and community, the topic might be sidelined, ignored or even refuted. Even for green-leaners, the flood, when it comes, is an excruciating conversion moment. In 2005, Carlisle was devastated by an unprecedented weather event, and I do mean unprecedented. Rain, unlike our usual January rain, did not let up. Snow did not act as it once might have, holding fast in the uplands; it fell as sleet and thawed rapidly. Then a huge storm arrived. The River Eden and the Petteril and the Caldew broke their banks, merging with a high tide in the Solway estuary. The resulting deluge drowned the city and much of it became impassable. The water level reached above the letterbox of my house; it was thirteen feet high in some places, with only a few roads surfacing as causeways. The situation was disaster-grade. That consoling phrase 'fifty-year flood' went out the window along with the storm surge, the free-flowing sewage, the plasterwork and people's valuables. Cue power outages and evacuations, helicopters rescuing the stranded from their rooftops, boats rowing above streets and carrying families and pets wrapped in blankets. The hazard itself was followed by a homelessness crisis, and an insurance crash. There were deaths, and extensive, long-lasting trauma. The weather system was not just a local front, of course; it had been cooked up by our shared international emissions, warmer temperatures and atmospheric volatility. Carlisle was a premonition of what will happen to our many riparian municipalities, and already is happening.

Such an event cannot pass a writer by unremarked – at least, not this writer. How could I ignore the sirens and the rising tide? I'd been reminded in the most shocking way that nature is not an exclusive rustic asset or confined to the countryside; it is everywhere, under and above us, surrounding society, permeating it, and because of human influence it is becoming unpredictable and perilous. I'd grown up with nature writing that was meditative, gentle or elegantly passionate; if it was politically charged, its campaigns were, like *Haweswater*'s, specific to a location and a philosophy, not multiply spearheaded and Earth-spanning. In literary terms, I needed to upscale, to expand the remit of environmental storytelling, retrofit the coordinates to make *where*'s plight matter universally.

My next nature novel, *The Carhullan Army*, would be set in the mountains of Cumbria again. It imagines a future landscape of increased precipitation, where the water table has risen to submerge towns and villages, disrupting civic life. Britain's government has become authoritarian, dependent on America for food production; population control limits female reproductive freedom; and resource wars are occurring around the globe. The Lake District is now an ungoverned zone – debatable land again. The protagonist, a woman called Sister, rejects the new social order and joins a group of off-grid eco-warriors, a group that has become, under the duress of its mercenary leader, paramilitary. Because under those circumstances, this is what I would do.

The pivot towards nightmarish societal collapse among nature's ruin seemed unavoidable. I knew now that the stakes were high, and the novel was a perfect way, through its immersive verisimilitude, to transport readers into the impending mess. *Carhullan* was reactive and it was designed to scare, to unseat readers. By novel three, and by witnessing catastrophe *in situ*, I was confident that Cumbria could be augmented via the filter of doomsday science fiction. Names warped, existing features reconstituted, another rotation of its identity. The rivers are unleashed – this time, the entire lowland is flooded, permanently. The old white-washed cottages are abandoned as residents are herded into shared urban accommodation. Across the abandoned fells, wilderness is creeping back, but it is uncanny, monobotanical. Eden is spoiled and runaway in ecological terms. What occurs in the story is brutal, as survivalism always is. It is the female characters who are at the steep end of societal breakdown in the novel. Gender, race, wealth – the impact of climate crisis is not felt equally.

I believed then that dystopia was an essential offering, a cultural tool; not merely presenting a confronting image of our world gone bad, but a way for readers to live out what was down the path. I tried, as I always do, to create a sensually experiential realm, to convey the horrors of new fascism, forced contraception and work duties, to present the extremes of self-sufficient upland living, and a volatile climate. In this way, readers might taste the future, and work through the most difficult moral choices about power, rules, agency and conflict. They might wake from the fugue state of headlong environmental disaster and want to re-route from its real-world track.

Now this sounds like a manifesto, more than a bit audacious. Novelists, particularly young female ones, aren't supposed to admit to high levels of ambition for their work – the hope of contributing something meaningful towards the prevention of planetary damage. But I'm from a valley where it wasn't unusual to see a woman with a rifle, operating big machinery, carrying an animal across her back. Living with nature is serious business, not for the faint-hearted, not a beauty pageant. And if the river raised me, the river in me must rise up when it's being fucked with, surely?

This is why the story offers an amplified vision of feminism, Mother Earth's soldiers. It's a loud shout about the dangers of an altering climate – a battle cry, really. Quite literally, it is the burning snowflake. *Carhullan* made the news beyond the books pages, landing on shelves around the time of more flooding in the UK and a looming economic crash, seeming to have its finger on the pulse. It was a gratifying retort to the assumption that rural novels haven't got any stones. One reviewer took issue with the research around guns and the mode of paramilitary training, while noting that I wrote fell-scapes very nicely. Essentially, a misogynistic nudge to *get back in your box, lady pastoralist*. I also took this as a win, the subversive disquiet created by the scenario of arming lasses in the pretty

Lakeland hills, of women on the offensive. To my mind, environmental fiction had to get tough – muscle up, buttercup; it had to exhibit dire current relevancy, and who better to lead the charge than a group of honed female uplanders.

I'm trying to justify my terms of engagement, and perhaps I don't have to. Maybe I'm overestimating fiction's heft. Aren't novels just 'flim-flam'? They don't influence politicians and policy, or lead to marches, riots and depositions. They're safe, inconsequential items, even if they pursue very naughty flights of fancy. In this country, writers aren't arrested for inflammatory, anti-establishment, even treasonous content. If a publisher is on board with an author's premise, there's no real risk, no skin in the game, as there is in countries where freedom of speech is curtailed and punishable. We are privileged. This is why our domestic fiction is perfectly positioned to play rough, to hold up the uglying mirror, to attempt a take-down of the ruling order; at the very least, it can create a provocative reverie, another realm of possibility. Fiction is a sanctioned device through which we may deliver revolution.

After creating a shitted-up little Eden, I found myself, for about a decade, undergoing a meditation about dystopia's qualities and purpose, its generativity. Many other bleak versions were written around this time, by writers all over the world, a catalogue of anxious *what-if* scenarios that featured desertification, mass migration, wildfires, floods, storms. Worst, by which I mean best, was Cormac McCarthy's *The Road*, its vastation of the biosphere and the planet's surface nearing total, its post-apocalyptic disintegration of civilisation so awful it is unconscionable. Dystopias are still popular, and they form the core of the genre now called climate fiction. Such stories undermine the reader's apathy, convoying ahead into destruction and disarray. They punish humanity for the many sins we have committed against the natural world. They do serve a purpose. Warning us, sickening and stupefying us, stripping us back to more primitive existences, or leading us into new regimes, hostile environments of our own making. But it's hard to know what impact dystopias have on people, whether they motivate political engagement and greener behaviour. Or whether they are simply like bad dreams, filling us with dread and hideous imagery, immersing us in hellscapes for the duration of their pages, only to dissipate when the book is closed, when we wake up in our lesser badlands. Do they subconsciously endorse our dark, destructive tendencies, and perpetuate Orwell's literary theory that utopias are unrenderable in fiction? Perhaps they are a maladaptive coping mechanism. It's important to remember that, for all the data and forecasting, Earth systems breakdown is likely to be extremely unpredictable. Psychologists note that not knowing an outcome can be more distressing than negative certainty: ambiguity is harder to cope with than pain. Dystopias, with their enacted and illustrative pessimism, might simply fill the anticipatory void, reinforcing our notions of inevitable doom, our failures, the fate of humanity to soil paradise.

But after The Fall, then what?

I felt that *Carhullan*'s final tenor was one of defiance and therefore victory against The System, so it wasn't purveying hopelessness. When reactivity becomes proactivity, there's impetus to fight the good fight. But its proposition felt like a limited contribution. Had I gone too far with the disturbing end-of-days vibe? Had I put readers off the subject? Had my dystopia simply been metabolised, and served the wrong purpose, in addition to pissing off the Cumbria tourism board? Climate crisis is such a vast and interconnected global issue, it's easy to feel isolated, frustrated, helpless or overwhelmed. My alternating novels have been about art – a nice secondment from which to observe the panorama and ponder its next season. I had ruined the neighbourhood and I wanted to mend it. When paradise is lost, which seeds should we plant in order to grow new gardens? Which stories should we tell in order to cast bolder, better projections of our habitats?

By the time I was ready to address environmental themes again, the artistic impetus had shifted. I'd been reading books on nature recovery and rewilding. I'd also been reading books about land ownership, the hierarchies of power, money and control that have existed for centuries in Britain and have curated its landscapes. A seismic political event was brewing – the Scottish independence referendum. With the border only a fast gallop from Cumbria, northerners were perhaps more attuned to the situation, and less facetious than those in Westminster. It wasn't uncommon to hear people in the county remark that if the Scots got a clear shot at it and the border was hoicked a little further south again, that might not be a bad thing. It was a time of very dynamic discussion about what freedom from the old

Union might look like, what sovereignty might mean for the economy, the Crown and of course the land mass, Scotland being uniquely positioned to sustain forms of wildlife the rest of the country cannot. Reformation does not come along often, certainly not brokered bloodlessly, as the referendum largely was, give or take a few loose punches in the pub. I couldn't help imagining the possibilities – a new precedent, for nationhood, society and landscape recovery. Previous to this, I'd also spent a lot of time in the Pacific Northwest, in Idaho and Washington state, where there were several notable species-reintroduction programmes – of apex predators. You see, I'd never stopped dreaming about those beautiful creatures in the enclosure downriver from my childhood home.

The grey wolf is a flashpoint in conservational and cultural terms. It was eradicated from Britain five hundred years ago, driven from its last known ranges in the Caledonian Forest. But it's still very much alive in our fairy tales and folklore, loping through our imaginations and our mythologies, pursued by fear, misinformation and bad press. At the risk of straying zoologically far from this lecture's remit, here are a few of its ecological pros. Wolves represent the pinnacle of ecosystems under which trophic cascade may lead to more equilibrium and web stability. Without such predators, environments become imbalanced and degraded, munched and mowed by sheep and deer; bringing them back can improve the health and biodiversity of landscapes right down to river life. Wolves could save huge amounts of money spent on estate culls; they could also generate large income from visitor centres (not a bad prospect for tourism in Alba). Grey wolves are extraordinary, matchlessly evolved creatures, anatomically superb – they are the most widely distributed terrestrial mammal on Earth, ranging from the Arctic to deserts. And they are native to Britain – they belong in its environments. They're hated too, across the world, by farmers, by politicians trying to appeal to farmers; in France under Charlemagne there was even an elite wolf-catching force set up, La Louveterie. Wolves haunt our nursery-book-stuck brains, terrorising the deep forests, gobbling up grandmothers and children. For the record – they don't.

Susiraja is a Finnish word, literally meaning 'wolf border' – the boundary between the capital region of Helsinki and the rest of the country. The term suggests that everything beyond this border is wilderness, including a more feral brand of people. As a Cumbrian savage, I interpret this as a compliment. Wolves don't abide by man-made boundaries, though, and for humans to live alongside them requires levels of respect, tolerance and compensation packages, a systematic understanding that we not only exist in enmeshed, cooperative habitats, we rely on them for planetary and atmospheric health, for agriculture and pollination, for homeostasis. It's interesting to compare wolves with our familial and societal power structures – other than the fact that male wolves actually do clean up the den site, we have more in common than you'd think. So this is what I wrote about next. I imagined a version of Britain that undergoes radical political change, where the idea of hereditary hierarchies and 'top dogs' is challenged, where the debate about rewilding is played out live and the lupine smear campaign is reversed. I gave Scotland its independence. (Novelists and the political God complex – discuss.) I let the wolf out of my mind, out of extinction and out of its cage, onto the pages.

The wolves of my childhood were kept in too small a confinement, and the wildlife park was shut in the 1980s after the introduction of new legislation, along with others that weren't suitable for their inhabitants. Until then, for every birthday excursion, I wanted to go and see them.

Side note: I was born in January. I'm a Twelfth Night baby, which might be why I favour the laws of Misrule. I have to say that seeing a wolf in the winter snow is the most astonishing, congruous thing imaginable, even when it's behind wire. The greys in Lowther Park were, essentially, domesticated; they were used to visitors and quite bored – they would stand and stare back at you. Those eyes – goldenish, or polar blue. The animals were so instinctually knowing and extrasensory in their assessment of you that they seemed almost clairvoyant. If you ran up and down outside their enclosure, sometimes they would run with you, mirroring, knowing your moves even before you did. Rachel Caine, the conservationist in *The Wolf Border*, does this as a child. She is my regenesis avatar; backed by an aristocrat with a huge estate and an unsavoury amount of money, she oversees the release of a breeding pair from quarantine into a self-sustaining enclosure, where they can hunt and breed. When the fence is sabotaged and they get out, she tracks them safely into the Scottish Highlands, where they are granted protection.

Fuck the red tape. What a heady venture this novel was! It felt experimental, beneficial and consequentialist, even if it curled the toes of Unionists and a conservative countryside brigade who would prefer not to share their hunting status. It trialled a set of divergent possibilities for a project still politically out of reach in Britain – one that is, nevertheless, feasible, with increasing popular support. Imagine a country under new dictates, with spectacular stretches of wilderness, converting itself into an ecological beacon.

I am not saying that *The Wolf Border* or *The Carhullan Army* represent my personal politics; they don't have to. I'm saying this is, in the most disruptive and thrilling way, what environmental novels can do. At a time when governments will not even endorse the simplest, cheapest measures to help nature – making swift bricks compulsory for new house building, for example – fiction can take the lead. It can recalibrate the whole operation, dismantling the infrastructures that perpetuate environmental harm, poor practices and anthropogenic impact, and allowing greener shoots to flourish. If we are ever to restore our landscapes and mitigate the worst damage, both nationally and internationally, we have to prosecute old models and believe in new ones; we have to be presented with examples, methods, new visions and positive stories. We must open the door to progressive possibilities. And once again, because the primary role of fiction is to induce ideas, to suspend disbelief and persuade the reader of inhabitance and occurrence, it is inherently capable of ecological conception, of strategic foresight, even blue-tomorrowing. No, fiction might not make events happen in the real world, and it certainly isn't obliged to create a perfect one, but it can promote alternatives, with what Octavia E. Butler described as 'an act of hope'.

Since *The Wolf Border*, I've sat on a number of nature-writing panels and taken part in debates about an author's role in the climate emergency. I believe, as I believed a quarter of a century ago, that writers choose and don't choose what to write about. They are inevitably prescribed by their cultural histories, provenance and agendas, but within these legacies they can find the basis for imaginative liberty. Writers don't have to tackle the climate emergency, but its significance becomes harder to steer away from. Whether it is the super-typhoon seen from the International Space Station in Samantha Harvey's *Orbital*, or the displacement of monarch butterflies in Barbara Kingsolver's *Flight Behavior*, the artistic needle is drawn to the subject. For me, the soul-searching has continued, about what I might add to the literary diet, and to the broader environmental conversation.

Fast-forward through the pandemic, when I wrote a novel about art and an even worse pandemic (no surprise that one didn't sell), to my last book, which, in this new era of climate justice, nature rights and the decentring of human dominion, lets an element narrate itself – literally. It's the story of Helm, the only named wind in Britain, a unique and spectacular entity that forms above – guess where – the Eden Valley. The wind tells its own life story, voices its own mythology, alongside the human tales told about it. Helm's lifespan is endangered, thanks to the species who named it; so too is humanity's. In a little observation hut nearby, Selima, a cloud pollution researcher, is collecting data from the skies around the Pennines. She is being hounded by climate sceptics, and she is heartbroken by the fate not only of Helm, but of Earth's vast natural treasury. Like the protagonists of my previous novels, she's defiant in her cause, she is Nature's heroine, no matter how small and lost she feels. The Helm, in reality, depends on several connected weather fronts and a stable maritime climate, a balance no longer guaranteed. If we know what wonders we are losing – killing off, actually – extraordinary phenomena all round the globe, and if we understand our inexorable relationship with the living world, does it make us less likely to sanction our destructive 'manifest destiny'? I don't know. But I keep making stories, crafting worlds that offer other possibilities. And I am not alone, trying to imagine systematic overhaul. Increasingly, novelists are turning their hand to solution stories and constructive virtualities, such as Kim Stanley Robinson's *The Ministry for the Future* and Courttia Newland's *A River Called Time*. These innovations shrug off restrictions and creative hindrances. It is exciting and chastening to think what the next generation of writers will produce, those who have read the Cassandra tales of their decedents, witnessed ineffectual policy and inaction; those who will suffer the results of ecological mismanagement, in very changed landscapes. All too often we hear it said that new generations will find ingenious answers to problems we've created and have failed to ameliorate. But that's just not good enough. The time is now, for the diverse ranks of nature writers to put their shoulders to the wheel, to pen new constitutions for the state of our environment, and to reach that ever-growing readership.

Even in my lifetime, people have begun saying we don't see much snow any more, not like when we were children. James Tate probably couldn't have known when he wrote 'Dear Reader' that his burning snowflake would take on sinister new meaning, that the degrees of global temperature rise measured since the poem's composition would take effect. Or that a brutal reversal might happen. I'm writing these lines after the morning news has featured updated scientific evidence about the likelihood of worst-case scenarios when the North Atlantic Current weakens, as sea water becomes desalinated from the melting poles, resulting in a new ice age for our little island. This, perhaps within decades. Imagine it. Imagine preventing it. Someone already is, I'm certain; not a feeble dreamer or an Orwellian doom-monger, but a Solar Punk. No. Nature writing is not just rhapsody, verses about daffodils and clouds; it's not about putting on stout boots and ascending the tallest mountain with a quill, *solus ambulo*, to remark upon beauty. Let's not forget, those writers were disrupters too, quite partial to a revolution. Nor is fiction flim-flam. It is a powerful collaboration of imagination between author and reader, a space of utmost resourcefulness and singularism, a safe haven stashed with creative explosives. If we choose to put it to purpose, environmental fiction is affirmative rebellion. It is a weapon, a hack – not greyware but greenware – the most persuasive device used to break inside the system, to reprogramme our narratives, and perhaps our belief in this world's options.

To end, I'd like to return home to the Lowther Valley and tell you about the river again. There's a different kind of light reflected on it now. Thanks to synergistic conservation work, the river's course has changed. Its headwaters have been meandered again, re-establishing gravel redds so Atlantic salmon can migrate upstream to breed, and reconnecting tens of hectares of floodplain. Water voles have been reintroduced, beavers are happily procreating downstream from my old house. Hundreds of thousands of trees have been planted, and species lost from the woodlands and commons have returned – golden eagles have even been sighted, undertaking reconnaissance, perhaps thinking of repopulating their eyrie above Haweswater. There's still traditional sheep farming, hefted flocks and hay meadows, but these practices are adapting. The landscape is on its way to restoration, a new era of biodiversity in the backyard, of which I could not be prouder. Nature is astonishing, uplifting, sublime and humbling; we know this, and those versions never get old. And when we imagine new, resurgent stories, sometimes they come to pass. There are no wolves yet, but I live, and write, in hope.